T0260002

SpringerBriefs in Applied Sciences and Technology

SpringerBriefs present concise summaries of cutting-edge research and practical applications across a wide spectrum of fields. Featuring compact volumes of 50–125 pages, the series covers a range of content from professional to academic.

Typical publications can be:

- A timely report of state-of-the art methods
- An introduction to or a manual for the application of mathematical or computer techniques
- A bridge between new research results, as published in journal articles
- A snapshot of a hot or emerging topic
- An in-depth case study
- A presentation of core concepts that students must understand in order to make independent contributions

SpringerBriefs are characterized by fast, global electronic dissemination, standard publishing contracts, standardized manuscript preparation and formatting guidelines, and expedited production schedules.

On the one hand, **SpringerBriefs in Applied Sciences and Technology** are devoted to the publication of fundamentals and applications within the different classical engineering disciplines as well as in interdisciplinary fields that recently emerged between these areas. On the other hand, as the boundary separating fundamental research and applied technology is more and more dissolving, this series is particularly open to trans-disciplinary topics between fundamental science and engineering.

Indexed by EI-Compendex, SCOPUS and Springerlink.

More information about this series at http://www.springer.com/series/8884

J. Jaidev Vyas · Balamurugan Gopalsamy
Harshavardhan Joshi

Electro-Hydraulic Actuation Systems

Design, Testing, Identification and Validation

 Springer

J. Jaidev Vyas
Mechanical Systems Design Group,
   Structures Division
CSIR-National Aerospace Laboratories
Bengaluru, Karnataka, India

Balamurugan Gopalsamy
CSIR-National Aerospace Laboratories
Bengaluru, Karnataka, India

Harshavardhan Joshi
CSIR-National Aerospace Laboratories
Bengaluru, Karnataka, India

ISSN 2191-530X   ISSN 2191-5318 (electronic)
SpringerBriefs in Applied Sciences and Technology
ISBN 978-981-13-2546-5   ISBN 978-981-13-2547-2   (eBook)
https://doi.org/10.1007/978-981-13-2547-2

Library of Congress Control Number: 2018954038

This Springer imprint is published by the registered company Springer Nature Singapore Pte Ltd.
The registered company address is: 152 Beach Road, #21-01/04 Gateway East, Singapore 189721,
Singapore

Preface

This brief presents the general guidelines for design, testing, identification and validation of electro-hydraulic actuation system. An analytical procedure for sizing the system for given loading and actuation rates has been elaborately explained. The system size is projected in terms of actuator capacity, servo valve flow rating, pump capacity and motor capacity. To verify the system design, the flow rates and system pressure values from experimental data have been compared with those from analytical calculations. The system performance in closed-loop operation is measured by comparing commanded position with the actual position. The flapper nozzle servo valve acting as direction control valve forms the core component of this system. As an attempt to provide a one-stop solution in completely understanding the working and dynamics of the valve, elaborate mathematical modelling is carried out from first principles and governing equations. However, due to OEM propriety, the servo valve internal parameters are not disclosed to the end-user. Thus, to identify the servo valve internal parameters, system identification approach has been adopted. A step-by-step approach for building mathematical models from experimental data using system identification toolbox (MATLAB) has been presented. The mathematical models built using first principles and through system identification are simulated, and the results are verified with the experimental results.

The brief is arranged in the following chapters: Chap. 1 lays out analytical approach to size the system in terms of load and flow rate calculations; Chap. 2 presents mathematical models of the flapper nozzle type direction control servo valve; Chap. 3 gives an insight into servo valve performance plots; Chap. 4 introduces the concept of system identification and lays out the procedure for identification and implementation; Chap. 5 presents simulation and test results followed by the conclusion in Chap. 6.

Bengaluru, India

J. Jaidev Vyas
Balamurugan Gopalsamy
Harshavardhan Joshi

Contents

About the Authors

J. Jaidev Vyas is a Scientist at the Mechanical Systems Design Group in the Structural Technologies Division of the National Aerospace Laboratories-CSIR, Bangalore, India. His research areas include aircraft hydraulic systems, fuel systems, ECS and landing gear. He received his M.Tech. in Thermal Engineering from the National Institute of Technology Warangal, India in 2008.

Balamurugan Gopalsamy is a Principal Scientist and Head of the Mechanical Systems Design Group, Structural Technologies Division at the National Aerospace Laboratories-CSIR, Bangalore, India. His research areas include aircraft hydraulic systems, fuel systems, ECS and landing gear systems. He completed his Ph.D. in Mechanical Engineering under the DAAD Sandwich programme at NIT, Durgapur, and RWTH Aachen-Fraunhofer IPT, Aachen in 2011.

Harshavardhan Joshi is a Senior Project Engineer at the Mechanical Systems Design Group in the Structural Technologies Division of the National Aerospace Laboratories-CSIR, Bangalore, India. His research areas include aircraft hydraulic systems. He received his B.E. in Mechanical Engineering from Sapthagiri College of Engineering, Bengaluru, India in 2014.

List of Figures

List of Tables

Chapter 1
Introduction

It is well-known fact that most of the smaller aircraft can be manually controlled since the loads on the flight control surfaces are less and well within the handling capability of pilot. However, to reduce the pilot's workload, hydraulic boosters are used in such aircraft with lower operating pressures. Redundancy of hydraulic system is not an essential consideration in such applications since the aircraft can be fully controlled by the pilot in the event of hydraulic power failures. However, many modern-day aircraft with large take-off weights and propelled by jet engines result in control loads which go beyond pilot's handling capability and positively calls for a powered flight control system. Further, this powered flight control system gets more complicated due to the interfacing of automatic flight control system (AFCS). Factors like reliability and high response characteristics with high operating load holding features required for the flight control operation have inadvertently resulted in the use of hydraulic systems in modern-day aircraft. These systems are used in conjunction with electrical interfaces like electrically operated direction control valves and sensors for precision actuation of the flight control surface. Hence, the system in whole is referred to as electro-hydraulic actuation systems.

Hydraulic fluid power is used in actuating flight controls and landing-gear systems due to its compact-sized equipment, high response rates and high load holding capabilities. Chapter 1 of this brief presents the design philosophy and a basic approach for preliminary sizing of aircraft hydraulic systems. For the purpose of clear understanding, design of a ground test rig, representative of typical electro-hydraulic actuation of flight control, surface has been presented. Few critical parameters like fluid selection, fluid contamination, temperature considerations, hydraulic line sizing, and troubleshooting, techniques are also covered.

It has been increasingly observed that, as a cost-cutting method, modern-day designers readily opt for mathematical modelling of the system and carry out

© The Author(s) 2019
J. J. Vyas et al., *Electro-Hydraulic Actuation Systems*, SpringerBriefs in Applied
Sciences and Technology, https://doi.org/10.1007/978-981-13-2547-2_1

rigorous simulations to realize a virtual prototype. This method is particularly beneficial as the exact behaviour of the physical system can be accurately predicted by computer simulations. In principle, the hydraulic power, from the centralized hydraulic pump, is routed via an electro-hydraulic servo valve into chambers of a double-acting actuator. The displacement of the actuator will result in generation of control forces and moments. The electro-hydraulic valves used for this purpose are mostly two staged flapper nozzle types. Chapter 2 presents a step-by-step approach to build mathematical models of the flapper nozzle servo valve from first principles.

However, though building mathematical models from first principles is advantageous, a thorough understanding of the physics of the system and expertise to identify and model, the system dynamics is required. This math-intensive method requires a very experienced control person to accomplish. It usually requires having good specifications for most if not all of the components. However, component information is often hard to obtain because of the OEM propriety. Therefore, the values of key system parameters are typically not calculated during the system design process.

A much easier method to build the mathematical model is through system identification. This process involves exciting the real system with a control input signal and recording the system response. The model unknown parameters (like natural frequency, damping ration in case of a servo valve) are estimated using an iterative trial-and-error techniques like predictor–corrector method, least squares method [1], etc., until the model response exactly emulates the real system response. Considering all the aforementioned disciplines, this brief intends to provide a holistic view of system design, system testing and system parameter identification in the context of electro-hydraulic actuation systems.

1.1 Design Philosophy

A simplified representation of electro-hydraulic actuation system for flight control is shown in Fig. 1.1. High-pressure fluid from the centralized pump is directed into either of the two chambers of the actuator. The actuator and the flight control surfaces are coupled through a crank (also called as moment arm). In real-time scenario, the pressure force developed by the actuator results in the development of torque about the hinge line of the flight control surface. This torque will encounter two major opposing loads: the load due to the inertia of the flight control surface and the aerodynamic loads [2, 3]. In order to simulate this scenario and check the overall system performance, a ground-based test rig is designed.

Applying force balance across actuator,

$$\text{Applied Force} = \text{Piston Inertial Loads} + \text{Viscous Damping Loads}$$
$$+ \text{FCS Inertial Loads} + \text{Aerodynamic Loads}$$

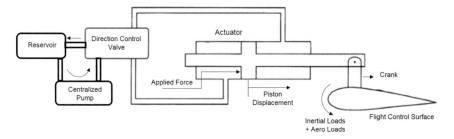

Fig. 1.1 Representation of electro-hydraulic actuation system

Neglecting the resistance due to piston inertial loads and the viscous damping loads,

$$\text{Applied Force} = \text{FCS Inertial Loads} + \text{Aerodynamic Loads}$$

In practice, simulating the aerodynamic loads on ground test rigs necessitates an opposing actuator to the primary flight actuator. However, for the sake of simplicity, the opposing actuator and hence the simulation of aerodynamic loads are being neglected in the present setup.

The system has been designed to actuate and deflect set of inertial discs according to the commanded profile. These discs represent inertial load on the actuator due to self-weight of the flight control surface. A simple crank arrangement (as shown in Fig. 1.1) has been adopted, and length of crank is fixed at 50 mm as initial design constraint. The linear reciprocating motion of the actuator will be converted to rotary deflection of the inertial discs through this arrangement. The actuator–inertial disc assembly is shown in Fig. 1.2. The system design has been carried out to deflect the inertial discs by ±30° at 3 Hz frequency.

Fig. 1.2 Actuator—Inertial discs assembly

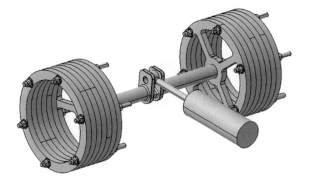

1.2   Design Prerequisites

First step in system design is to define the primary parameters like system fluid pressure, maximum fluid flow rate requirements and the hydraulic fluid that needs to be used. Defining these three parameters is very much essential as a part of initial design and they cannot be easily changed during detail design stage.

For a defined power level, considering lower system pressure will increase the size of the system and opting for high system pressure will necessitate increased wall thickness. Hence, selected system pressure has a bearing on the hydraulic system weight in meeting the required power levels. It is interesting to note that the hydraulic system weight for a defined power level will considerably increase if the defined fluid pressure is below 3000 psi and above 4500 psi. Hence, it is preferable to develop the hydraulic system for fluid pressure range 3000–4500 psi and advantage of weight reduction. Further, it is preferable to use standard pressures as used in the industry than defining our own system fluid pressure. This will ensure ready availability of various system components from the industry.

Another important parameter to be defined in the system is the system flow rate requirements. It is worthy here to note that, while system pressure will depend on the resistive loads on actuator, the system flow rate requirement will depend on the desired actuation rate. The flow rate requirement will determine the servo valve capacity, pump capacity, and the motor capacity.

1.3   Preliminary Design

As mentioned earlier, test rig is designed to actuate static inertial discs representing the self-weight of the flight control surface. The actuator will deflect the inertial discs (having max MoI = 2.68 Kg-m^2) by ±30° from mean position at a maximum frequency of 3 Hz. Succeeding section outlines the methodology adopted in determining the load on the actuator, actuator sizing, and sizing of the centralized pump and motor.

1.3.1   Actuator Load Calculations

The actuator and the inertial discs are connected to each other through a simple crank mechanism as shown in Fig. 1.2. Neglecting the resistive forces due to damping and spring stiffness, the torque balance equation across this assembly can be written as

$$T = F * rCos\theta = I * \alpha \qquad (1.1)$$

where F is the load on the actuator, I is the net moment of inertia of the inertial discs, α is the angular acceleration, r is the crank length and the term r*cos θ represents the effective crank length for maximum deflection. From Eq. (1.1),

$$F = \frac{I.\alpha}{r * \cos\theta} \tag{1.2}$$

The commanded angular deflection profile can be represented by

$$\theta = \frac{\pi}{6} * \sin(2 * \pi * f * t) \tag{1.3}$$

where f is the operating frequency.

Differentiating Eq. (1.3), angular velocity profile can be represented by

$$\dot{\theta} = \omega = \frac{\pi}{6} * (2 * \pi * f) * \cos(2 * \pi * f * t) \tag{1.4}$$

Differentiating Eq. (1.4), angular acceleration of the motion can be represented by

$$\ddot{\theta} = \alpha = \frac{\pi}{6} * (2 * \pi * f)^2 * \sin(2 * \pi * f * t) \tag{1.5}$$

Now, substituting f = 3 Hz and applying maximum angular velocity and maximum angular acceleration conditions to Eqs. (1.4) and (1.5),

$$\dot{\theta}_{max} = \omega_{max} = \frac{\pi}{6} * (2 * \pi * f) * [\cos(2 * \pi * f * t)]_{max}$$
$$= 9.8696 \text{ rad/s} \tag{1.6}$$

$$\ddot{\theta}_{max} = \alpha_{max} = \frac{\pi}{6} * (2 * \pi * f)^2 * [\sin(2 * \pi * f * t)]_{max}$$
$$= 186 \text{ rad/s}^2 \tag{1.7}$$

Thus, for the application under consideration, the discs will deflect at maximum angular velocity of 9.86 rad/sec and maximum angular acceleration of 186 rad/sec^2.

1.3.2 Maximum Loading Condition

Moment of inertia (of 14 discs) equals 2.68 kg-m^2, Crank radius is fixed at r = 0.05 m, maximum deflection $\theta_{max} = \pm 30°$ and maximum angular acceleration, $\alpha_{max} = 186.037$ rad/sec^2 as calculated above. From Eq. (1.2),

$$F_{max} = \frac{I * \alpha_{max}}{r * \cos\theta_{max}} = \frac{2.68 * 186.037}{0.05 * \cos(30°)} = 11510 \text{ N} \tag{1.8}$$

Considering a FOS of 1.5, the maximum permissible load on actuator = 1.5 *11510 = 17265 kN. Thus for a fixed supply pressure of 220 bar, the actuator area can be calculated using

$$A = \frac{F}{P} = \frac{17265}{22000000} = 0.0007658 \quad m^2$$

Calculation of maximum pressure differential 'ΔP' across actuator:

$$F_{max} = (\Delta P)_{max} * A \tag{1.9}$$

$$(\Delta P)_{max} = \frac{F_{max}}{A} = 150.36 \text{ bar}$$

Following the similar procedure, analytical calculations are carried out for three test cases (each carrying 18 different sets with varying load and varying frequency) has been tabulated below (Tables 1.1, 1.2 and 1.3).

Table 1.1 Analytically calculated actuator pressure forces for commanded angle of 10°

Disc sets	Net inertia (Kg-m^2)	Frequency					
		1 Hz		2 Hz		3 Hz	
		Max force (kN)	Max delta P (bar)	Max force (kN)	Max delta P (bar)	Max force (kN)	Max delta P (bar)
1 sets (2 discs)	0.38	0.05	0.70	0.21	2.80	0.48	6.30
2 sets (4 discs)	0.77	0.11	1.40	0.43	5.60	0.96	12.59
3 sets (6 discs)	1.15	0.16	2.10	0.64	8.39	1.45	18.89
4 sets (8 discs)	1.53	0.21	2.80	0.86	11.19	1.93	25.18
5 sets (10 discs)	1.91	0.27	3.50	1.07	13.99	2.41	31.48
6 sets (12 discs)	2.30	0.32	4.20	1.29	16.79	2.89	37.78
7 sets (14 discs)	2.68	0.38	4.90	1.50	19.59	3.38	44.07

Table 1.2 Analytically calculated actuator pressure forces for commanded angle of 20°

Disc sets	Net inertia (Kg-m^2)	Frequency					
		1 Hz		2 Hz		3 Hz	
		Max force (kN)	Max delta P (bar)	Max force (kN)	Max delta P (bar)	Max force (kN)	Max delta P (bar)
1 set (2 discs)	0.38	0.11	1.47	0.45	5.87	1.01	13.20
2 sets (4 discs)	0.77	0.22	2.93	0.90	11.73	2.02	26.39
3 sets (6 discs)	1.15	0.34	4.40	1.35	17.60	3.03	39.59
4 sets (8 discs)	1.53	0.45	5.87	1.80	23.46	4.04	52.79
5 sets (10 discs)	1.91	0.56	7.33	2.25	29.33	5.05	65.98
6 sets (12 discs)	2.30	0.67	8.80	2.70	35.19	6.06	79.18
7 sets (14 discs)	2.68	0.79	10.26	3.14	41.06	7.07	92.38

Table 1.3 Analytically calculated actuator pressure forces for commanded angle of 30°

Disc sets	Net inertia $(Kg\text{-}m^2)$	Frequency					
		1 Hz		2 Hz		3 Hz	
		Max force (kN)	Max delta P (bar)	Max force (kN)	Max delta P (bar)	Max force (kN)	Max delta P (bar)
1 set (2 discs)	0.38	0.18	2.39	0.73	9.55	1.64	21.48
2 sets (4 discs)	0.77	0.37	4.77	1.46	19.09	3.29	42.96
3 sets (6 discs)	1.15	0.55	7.16	2.19	28.64	4.93	64.44
4 sets (8 discs)	1.53	0.73	9.55	2.92	38.19	6.58	85.92
5 sets (10 discs)	1.91	0.91	11.93	3.66	47.73	8.22	107.40
6 sets (12 discs)	2.30	1.10	14.32	4.39	57.28	9.87	128.88
7 sets (14 discs)	2.68	1.28	16.71	5.12	66.82	11.51	150.36

1.3.3   Flow Rate Requirement Calculation

The flow rate required to deflect the inertial load by ± 30° at 3 Hz is calculated by applying flow continuity equation. The maximum linear velocity V_{max} of the piston is calculated using

$$V_{max} = r * \omega_{max}$$
$$= 0.05 * 9.8696 = 0.463 \text{ m/s} \tag{1.10}$$

By applying continuity equation, we have the maximum flow rate requirement to be

$$Q_{max} = A * V_{max}$$
$$= 0.0007658 * 0.493$$
$$= 0.0003775 \text{ m}^3/s \tag{1.11}$$
$$Q_{max} = 22.65 \text{ LPM}$$

Therefore, to suffice this requirement, a 25 LPM pump and 25 LPM servo valve are chosen. It has to be noted that the flow rate requirement is dependent on the frequency of operation and is independent of the load. For different operating frequencies, the flow rate calculations have been tabulated below (Tables 1.4, 1.5 and 1.6).

1.3.4   Motor Selection

The theoretical calculation of the power of an electric motor required is shown below:

Table 1.4 Analytically calculated actuator flow rates for commanded angle of 10°

Frequency (Hz)	Max angular velocity (rad/sec)	Max linear velocity (m/sec)	Max flow rate (LPM)
1	1.10	0.05	2.52
2	2.19	0.11	5.04
3	3.29	0.16	7.56

Table 1.5 Analytically calculated actuator flow rates for commanded angle of 20°

Frequency (Hz)	Max angular velocity (rad/sec)	Max linear velocity (m/sec)	Max flow rate (LPM)
1	2.19	0.11	5.04
2	4.39	0.22	10.08
3	6.58	0.33	15.12

Table 1.6 Analytically calculated actuator flow rates for commanded angle of 30°

Frequency (Hz)	Max angular velocity (rad/sec)	Max linear velocity (m/sec)	Max flow rate (LPM)
1	3.29	0.16	7.56
2	6.58	0.33	15.12
3	9.87	0.49	22.65

$$\text{Power}, P = \frac{\text{Flow Rate(GPM)} * \text{Pressure(psi)}}{1714} \qquad (1.12)$$

$$P = \frac{6.6050 * 3200}{1714} = 12.33 \text{ HP}$$

The maximum motor power can be approximated as 15 HP. The major LRU details have been mentioned in Table 1.7.

1.3.5   Fluid Selection

Mainly three types of hydraulic fluids are used in aerospace applications. In Type I systems, Mil-H-5606 is used. It is a mineral-based hydraulic fluid with operating temperature from −40 to +100 °C. In extreme cases, the fluid can, however, withstand temperature up to +135 °C. In Type II systems, Mil-H-83282 fluid is used. It has high fire-resisting properties and can withstand temperatures beyond +100 °C. This class of fluid is specially used for machines operating in areas

Table 1.7 Test rig LRU details

Sl No	LRU	Make	Capacity	Quantity
1	Actuator	Servo controls	1T	1
2	Axial piston pump	Veljan	25 LPM @ 206 bar	1
3	AC motor	Seimens	11 kW, 1445 RPM	1
4	Flapper nozzle valve	Servo controls	25 LMP	1
5	Diaphragm-type pressure transducer	Wika	400 bar	3
7	Turbine-type flow meters	VSE	0.03–40 LPM 250 bar	3
8	Gas-charged accumulator	Olear	2.5 L	1
9	High-pressure filter	MP Filtri	30 LPM 250 bar, 10 μ	1
10	Return line filter	MP Filtri	50 bar, 15 μ	2
11	Position sensor (LVDT)	MTS	–	1
12	Load cell	Interface	12.5 kN	1

susceptible to fire. The major drawback with MIL-H-83282 fluid is that it becomes highly viscous below –20 °C. Skydrol is the other hydraulic fluid which is mainly used in commercial aircraft. Considering all the above-mentioned aspects, Mil-H-5606 is preferred for the system under consideration.

1.3.6 Design Summary

The powered hydraulic fluid is supplied by an axial piston pump at 25 LPM and 220 bar supply pressure. A double-acting hydraulic linear actuator with net effective area of 0.0007658 m^2 and the stroke equal to 70 mm is used. The actuator is fixed horizontally to a test bed. The modelled current input to the servo amplifier is given through a 12-bit RMC 70 controller. All the signals from the sensors are recorded by an NI DAQ card through 16-bit analogue-to-digital converter. Figure 1.3 gives the schematic of the experimental setup followed by pictures of the test rig in Fig. 1.4 (Table 1.8).

1.4 Verification of System Performance

To verify the system performance, sinusoidal deflection command of $\pm 10°$, $\pm 20°$ and $\pm 30°$ at different operating frequencies is commanded. The experimentally determined actuator flow rates and loads on the actuator are compared with analytically calculated values in each test case (Fig. 1.5 and Table 1.9).

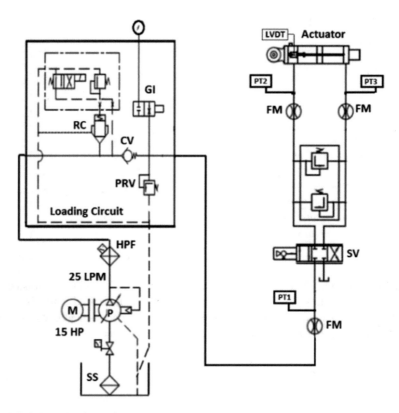

Fig. 1.3 Schematic of experimental setup

Case 2: ±20° sinusoidal command, six discs (Fig. 1.6 and Table 1.10)
Operating Frequency: 2 Hz (Fig. 1.7 and Table 1.11)
Operating Frequency: 3 Hz (Fig. 1.8 and Table 1.12)
Case 3: ±30° sinusoidal command, 10 discs (Fig. 1.9 and Table 1.13)
Operating Frequency: 1 Hz (Fig. 1.10 and Table 1.14)
Operating Frequency: 2 Hz (Fig. 1.11 and Table 1.15)

In each of the above test cases, a close match between commanded deflection and actual deflection can be observed. There is no overshoot and phase difference is also negligible. This indicates fine performance of the system aided by the controller. In practice, the actual deflection and commanded deflections are instantaneously compared on the controller's GUI and values of proportional gain, and integral gains are varied until desired profile is realized. Additionally, most controllers come with provisions for adjusting velocity feedforward and acceleration feedforward gains. These are open-loop gains commanded to the controller to overcome the system nonlinearity.

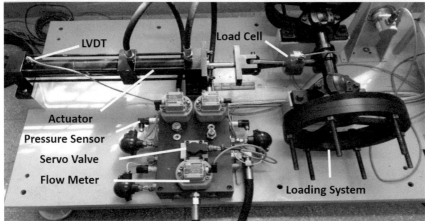

Fig. 1.4 Experimental system

Table 1.8 List of abbreviations

SS	Suction strainer
HPF	High-pressure filter
RC	Relief cartridge
CV	Check valve
GI	Gauge isolator
PRV	Pressure relief valve
PT	Pressure transducer
FM	Flow meter
SV	Servo valve

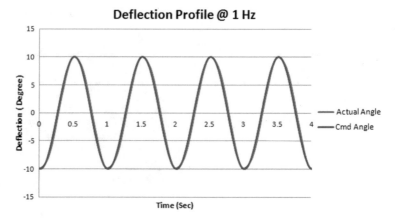

Fig. 1.5 Sinusoidal command of 1 Hz (4 discs)

Table 1.9 Comparison of results for commanded angle of ± 10° and operating frequency of 1 Hz (4 discs)

% Pos error	Actuator load (Analytical)	Actuator load (Test)	Flow rate (Analytical)	Flow rate (Test)
0.513	0.11 kN	0.3 kN	2.52 LPM	2.3 LPM

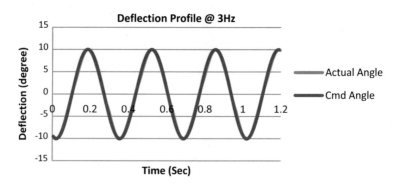

Fig. 1.6 Sinusoidal command of 3 Hz (4 discs)

Table 1.10 Comparison of test results for commanded angle of ± 10° and operating frequency of 3 Hz (4 discs)

% Pos error	Actuator load (Analytical)	Actuator load (Test)	Flow rate (Analytical)	Flow rate (Test)
0.2	0.96 kN	1.18 kN	7.56 LPM	7.10 LPM

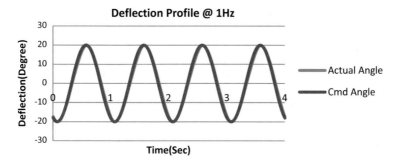

Fig. 1.7 Sinusoidal command of 1 Hz (6 discs)

Table 1.11 Comparison of test results for commanded angle of ± 20° and operating frequency of 1 Hz (6 discs)

% Pos error	Actuator load (Analytical)	Actuator load (Test)	Flow rate (Analytical)	Flow rate (Test)
0.96	0.34 kN	0.36 kN	5.04 LPM	5.30 LPM

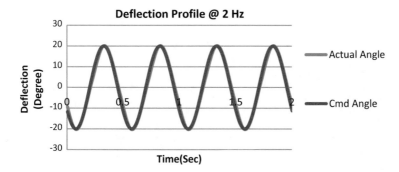

Fig. 1.8 Sinusoidal command of 2 Hz (6 discs)

Table 1.12 Comparison of test results for commanded angle of ± 20° and operating frequency of 2 Hz (6 discs)

% Pos error	Actuator load (Analytical)	Actuator load (Test)	Flow rate (Analytical)	Flow rate (Test)
0.96	1.35 kN	1.7 kN	10.08 LPM	10.43 LPM

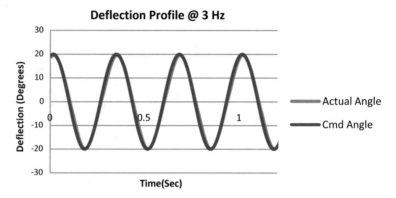

Fig. 1.9 Sinusoidal command of 3 Hz (6 discs)

Table 1.13 Comparison of test results for commanded angle of ± 20° and operating frequency of 3 Hz (6 discs)

% Pos error	Actuator load (Analytical)	Actuator load (Test)	Flow rate (Analytical)	Flow rate (Test)
1.39	3.03 kN	3.09 kN	15.12 LPM	15.5 LPM

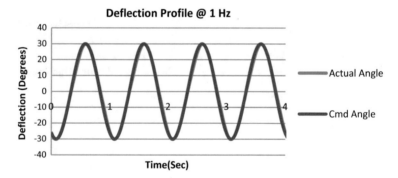

Fig. 1.10 Sinusoidal command of 1 Hz (10 discs)

Table 1.14 Comparison of test results for commanded angle of ± 30° and operating frequency of 1 Hz (10 discs)

% Pos error	Actuator load (Analytical)	Actuator load (Test)	Flow rate (Analytical)	Flow rate (Test)
0.64	0.91 kN	1.1 kN	7.56 LPM	7.82 LPM

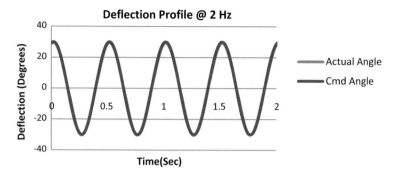

Fig. 1.11 Sinusoidal command of 2 Hz (10 discs)

Table 1.15 Comparison of test results for commanded angle of $\pm$ 30° and operating frequency of 2 Hz (10 discs)

% Pos error	Actuator load (Analytical)	Actuator load (Test)	Flow rate (Analytical)	Flow rate (Test)
1.57	3.66 kN	3.88 kN	15.12 LPM	15.87 LPM

1.5 Hydraulic Line Sizing

Hydraulic lines form important part of hydraulic system and inter-connect various hydraulic equipment. Both rigid pipelines and flexible hoses will be used in the existing test rig. Flexible hoses are generally heavier than the equivalent rigid pipes. Hence, flexible hoses are specifically used where hydraulic port of the equipment has a relative movement w.r.t. structure or associated hydraulic equipment. Another important task is to define size of the hydraulic lines inter-connecting various hydraulic equipment. For this, the maximum expected fluid flow in each of the hydraulic lines is computed. Then using the sizing thumb rule, the lines are sized. As per the sizing thumb rule, the high-pressure lines carrying fluid under pressure above 2000 psi shall not exceed a fluid flow velocity more than 5 m/s, low-pressure lines carrying fluid below 100 psi shall not exceed a fluid flow velocity more than 3 m/s and pump suction lines shall not exceed a fluid flow velocity more than 1 m/sec. In many of the existing aerospace applications, stainless steel tubes are used for high-pressure lines and aluminium alloy tubes are used for low-pressure lines. Now, the trend is to use titanium alloy tubes both for the high- and low-pressure lines due to its low density, higher fatigue life and better flexibility. Soft tubing will increase in weight, and hard tubing reduces flexibility. Hence, a balanced combination will work out as the best solution.

1.6   Contamination Control

Even with a good awareness on prevention and control of contamination level in hydraulic systems, it is established that 70% of hydraulic system failures are due to fluid contamination.

Hence, proper care of hydraulic fluid as stored, handled and used in hydraulic system will greatly improve system performance and component life. Invariably filters need to be used in the hydraulic system to control contamination level of the system during its service. Filter types and location are of great importance. In addition to main pressure and return line filters, inlet strainers are used in pump suction line. Filters are also named as full flow and by-pass filters depending on their location in a hydraulic system. Full flow in-line filters are generally fitted with a by-pass valve. Filters are designated by their micron size (size of the particle which will be trapped by the filter element) and beta ratio (ratio of existence of a defined particle size between upstream and downstream fluids across a filter element).

NAS 1638 standard is generally used to define the hydraulic fluid contamination level by quantity of particles present in the fluid in aerospace hydraulic systems. In general, the overall contamination in aerospace hydraulic systems is limited to NAS 1638 Class-8 and system must function satisfactorily up to NAS 1638 Class-10 (Table 1.16).

Table 1.16   15 NAS 1638 fluid contamination chart

NSA 1639 Fluid contamination class	Number of particles per 100 ml μ range and particle size (in microns)				
	5–15	15–25	25–50	50–100	>100
00	125	22	4	1	0
0	250	44	8	2	0
1	500	88	16	3	1
2	1,000	178	32	6	1
3	2,000	356	63	11	2
4	4,000	712	126	22	4
5	8,000	1,425	253	45	8
6	16,000	2,800	506	90	16
7	32,000	5,700	1,012	180	32
8	64,000	11,400	2,000	360	64
9	128,000	22,800	4,100	720	128
10	256,000	45,600	8,100	1,440	256
11	512,000	91,200	16,200	2,800	512
12	1,000,000	182,000	32,400	5,800	1,024

References

1. M. Jelali, A. Kroll, *Hydraulic Servo Systems–Modelling, Identification and Control* (Springer, 2003)
2. K. Kang, M. Patcher, C.H. Houpis, S. Rasmusssen, Modelling and control of electro hydrostatic actuator (Department of Electrical and Computer Engineering, Airforce Institute of Technology (AFIT/ENG) Wright Patterson AFB OH 45433, 1994)
3. S. Frischemeir, *Electro-hydrostatic Actuators for Aircraft Primary Flight Control*. Technical University Hamburg-Harburg

Chapter 2
Mathematical Modeling of Flapper Nozzle Valve

In this section, mathematical models for the flapper nozzle type servo valve are presented. These models are derived from the fundamental governing equations and are important in the context of understanding the dynamics of the system and realizing a virtual prototype for simulations. The various dynamic phenomenon involved in the system are represented using ordinary differential equations. These differential equations are solved using Laplace Transform technique to build single input single output system models in time domain. The transfer function model thus presented represents the spool stage and the flapper stage respectively and provides a good tool to understand system behaviour under various operating frequencies and input command of various amplitudes and patterns.

2.1 Flapper Nozzle Valve Flow Analysis

Let us consider a four-way spool valve as shown in the Fig. 2.1. The numbers at the ports refer to the subscripts of the flow and the area of the ports.

Let the spool be given a positive direction x_v from the null position that is $x_v = 0$. This is chosen to be the symmetrical position of the spool in the sleeve. As we are only interested in the steady state characteristics, the compressibility flows are zero and the continuity equation for the two valve chambers are

$$Q_L = Q_1 - Q_4 \tag{2.1}$$

$$Q_L = Q_3 - Q_2 \tag{2.2}$$

$$P_L = P_1 - P_2 \tag{2.3}$$

© The Author(s) 2019
J. J. Vyas et al., *Electro-Hydraulic Actuation Systems*, SpringerBriefs in Applied Sciences and Technology, https://doi.org/10.1007/978-981-13-2547-2_2

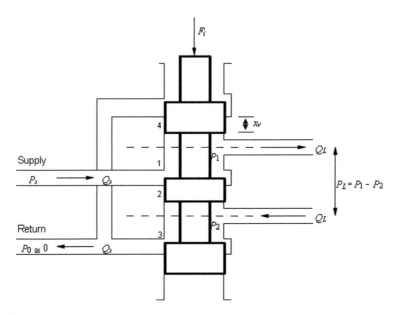

Fig. 2.1 A Typical three land four way spool valve

where, Q_L is the flow through the load and P_L is the pressure drop across the load. Flows through valve orifices are described by the flow through the orifices equation. The general equation for flow through orifice is given by,

$$Q_L = C_d A \sqrt{\frac{1}{\rho}} \sqrt{P_S - \frac{x_v}{|x_v|}} \tag{2.4}$$

where,
   C_d is the coefficient of discharge.
   A is the orifice area.
   P_s is the supply pressure.
   The flow rate across the load is equal to the product of the area and square root of the difference of the supply pressure and load pressure, which again is multiplied by inverse of the square root of the density of the fluid used and the coefficient of discharge.
   Applying Eq. (2.4) to all the four orifices,

$$Q_1 = C_d A_1 \sqrt{\frac{2(P_s - P_1)}{\rho}} \tag{2.5}$$

$$Q_2 = C_d A_2 \sqrt{\frac{2(P_s - P_2)}{\rho}} \tag{2.6}$$

$$Q_3 = C_d A_3 \sqrt{\frac{2(P_2 - P_o)}{\rho}} \tag{2.7}$$

$$Q_4 = C_d A_4 \sqrt{\frac{2(P_1 - P_o)}{\rho}} \tag{2.8}$$

Q1, Q2, Q3, Q4 and A_1, A2, A3, A4 are the flow through their respective orifice and corresponding orifice areas. The return line P_o is neglected as the value is very much lesser than the other pressure than the other pressures involved. P_1, P_2 are the actuator chamber pressure. The orifice area depends on the valve geometry and the valve opening depends on the spool displacement, so the four areas can be defined as:

$$A_1 = A_1(x_v) \quad A_2 = A_2(x_v) \quad A_3 = A_3(-x_v) \quad A_4 = A_4(-x_v) \tag{2.9}$$

From the above equations, we can infer that when the equations are solved simultaneously they yield load flow as a function of valve position and load pressure.

$$Q_L = Q_L(x_v, P_L) \tag{2.10}$$

In vast majority of cases, the valve orifices are matched and symmetrical. Matched orifices require that,

$$A_1 = A_3 \tag{2.11}$$

$$A_2 = A_4 \tag{2.12}$$

Symmetrical orifices require that,

$$A_1(x_v) = A_2(-x_v) \tag{2.13}$$

$$A_3(x_v) = A_4(-x_v) \tag{2.14}$$

At neutral position of the spool, all four orifices are equal

$$A_1(0) = A_2(0) \approx A_n \tag{2.15}$$

From the above equations, we can conclude that one orifice area needs to be defined because the other areas follow from it. The orifice areas are for linear with valve stroke, and then the only defining factor would be the slot opening or width of the slot in the valve sleeve 'w'. 'w' is also called the area gradient and had units $(length)^2/(length)$. When the orifices are matched and symmetrical,

$$Q_1 = Q_3 \tag{2.16}$$

$$Q_2 = Q_4 \tag{2.17}$$

Substituting Eqs. (2.5), (2.7), (2.11) into (2.16) yields,

$$P_s = P_1 + P_2 \tag{2.18}$$

From Eq. (2.3) we have PL = P_1-P_2
Solving Eqs. (2.3) and (2.18),

$$P_1 = \frac{P_s + P_L}{2} \tag{2.19}$$

$$P_2 = \frac{P_s - P_L}{2} \tag{2.20}$$

From the above two equations, we can understand that for a matched and symmetrical valve with no load that is $P_L = 0$, the pressure in each motor line is $0.5P_s$.

The total supply flow can be written as,

$$Q_s = Q_1 + Q_2 \tag{2.21}$$

$$Q_s = Q_3 + Q_4 \tag{2.22}$$

Using the Eqs. (2.16), (2.17), (2.18), (2.19), (2.20) in (2.1) and (2.2).

$$Q_L = C_d A_1 \sqrt{\frac{P_s - P_L}{\rho}} - C_d \, A_2 \sqrt{\frac{P_s - P_L}{\rho}} \tag{2.23}$$

$$Q_s = C_d A_1 \sqrt{\frac{P_s - P_L}{\rho}} + C_d \, A_2 \sqrt{\frac{P_s - P_L}{\rho}} \tag{2.24}$$

From the Eq. (2.10)

$$Q_L = Q_L(x_v, P_L)$$

We can express this function as a Taylor's series about a particular point $Q_L = Q_{Ln}$

$$Q_L = Q_{L1} + \frac{\delta Q_L}{\delta x_v}\bigg|_1 \Delta x_v + \frac{\delta Q_L}{\delta P_L}\bigg|_1 \Delta P_L + \cdots$$

If we confine to the vicinity of the operating point, higher order infinitesimals are negligibly small and it becomes.

$$Q_L - Q_{L1} = \Delta Q_{L1} \approx \frac{\delta Q_L}{\delta x_v}\bigg|_1 \Delta x_v + \frac{\delta Q_L}{\delta P_L}\bigg|_1 \Delta P_L + \qquad (2.25)$$

The partial derivatives required are obtained by differentiation of the equation for the pressure flow curves. These partials define the two most important parameters for a valve. The flow gain is defined by

$$K_q = \frac{\partial Q_L}{\partial x_v} \qquad (2.26)$$

K_q, is the flow gain, which relates flow through the load to the valve opening when pressure remains steady.

The flow pressure coefficient is defined by

$$K_c = -\frac{\partial Q_L}{\partial P_L} \qquad (2.27)$$

K_c, Pressure flow coefficient, relates flow through the load to pressure when the valve opening remains steady. Using Eqs. (2.26), (2.27) in (2.25)

$$\Delta Q_L = K_q \, x_v - K_c \, P_L \qquad (2.28)$$

2.2   Flow Forces

Hydraulic spool valves are used for controlling hydraulic power for a wide number of mechanical engineering applications. In order for the hydraulic spool valve to accomplish its objective, its position must be controlled by a spool-valve actuation device. As the actuation device attempts to move the spool, it must overcome the forces that act on the spool valve, which result from the momentum of the fluid passing through the valve itself. These momentum effects are familiarly known as flow forces and are generally quite substantial when a large amount of flow is passing through the valve. The flow forces that act on the spool valve have both transient and steady components [4].

2.2.1 *Flow Forces on Spool Valves*

Flow forces are also called flow induced forces, Bernoulli's forces or hydraulic reaction forces. These names refer to those forces acting on a valve as a result of fluid flowing in the valve chambers and through the valve orifices. When Newton's second law is applied to the valve in terms of flow rate, we get,
  Force = mass × acceleration

$$F \,=\, ma \tag{2.29}$$

Mass = density × volume

$$m \,=\, \rho V \tag{2.30}$$

Acceleration = velocity/time
Velocity = mass flow rate/Area
Acceleration = mass flow rate/(Area × time)
Time = mass flow rate/volume
a = [Mass flow rate/Area] × [Mass flow rate/volume]
a = [Mass flow rate]^2/[Area × Volume]

$$a = \frac{Q^2}{A \text{ x } V} \tag{2.31}$$

$$F = \frac{\rho \, Q^2}{A} \tag{2.32}$$

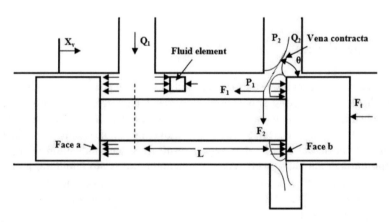

Fig. 2.2 Flow forces on the spool valve due to flow leaving the chamber

where,

Q$_1$, Q$_2$ are the flow rate through the respective orifices.

x$_v$ is the spool displacement.

A$_1$, A$_2$ areas at inlet and outlet orifices.

For the above Fig. 2.2, Eq. (2.32) can be written as,

$$F_j = \frac{\rho Q_2^2}{A_2} \tag{2.33}$$

$$F_j = \frac{\rho Q_2^2}{C_c A_o} \tag{2.34}$$

where,

C$_c$ is the contraction coefficient.

F$_j$ is the force of the jet.

A$_0$ is the area of the orifice, = w x$_v$.

From Newton's third law, this jet force has an equal and opposite reaction force which can be resolved into the horizontal and vertical component as the force acts at an angle.

Horizontal component,

$$F_1 = -\ F_j \cos\theta \tag{2.35}$$

Vertical component,

$$F_2 = -\ F_j \sin\theta \tag{2.36}$$

The lateral or the vertical component tends to push the valve spool sideways against the sleeve and cause sticking. The axial force acts in a direction to close the valve port. Assuming that fluid is incompressible and using the continuity equation along with the orifice flow equation, we get,

$$Q_1 = Q_2 = C_d\ A_o \sqrt{\frac{2(P_1 - P_2)}{\rho}} \tag{2.37}$$

$$C_d = C_v \text{x}\ C_c \tag{2.38}$$

$$A_o = w\ x_v \tag{2.39}$$

where,

C$_v$ is the coefficient of velocity.

P$_1$ and P$_2$, pressures at their respective sections.

Using Eq. (2.37) in (2.34)

$$F_j = 2\,C_dC_vA_o(P_1 - P_2) \tag{2.40}$$

From Eq. (2.35)

$$F_1 = -2C_c(C_v)^2A_o(P_1 - P_2)\cos\theta \tag{2.41}$$

Steady state axial flow force equation is given by Eq. (2.41).

When there is no radial clearance between the valve spool and sleeve, one can use the following data from [3],

$$C_d = 0.61 \quad C_v = 0.98 \quad \theta = 69° \quad \cos\theta = 0.358$$

$$F1 = 0.43\Delta P \text{ w } x_v \tag{2.42}$$

When the mass of the fluid is accelerated then a force is produced, the force produced reacts on the face of the spool valve lands. The magnitude of transient flow force is given by Newton's second law,

$$F = m \text{ x } a$$
$$F = \rho\,V\,\frac{d(Q_1/A_v)}{dt}$$
$$F = \rho\,A_v\,L\,\frac{d(Q_1/A_v)}{dt} \tag{2.43}$$
$$F = \rho\,L\,\frac{d(Q_1)}{dt}$$

where,

L is the damping length.

From Eq. (2.37) we have,

$$Q_1 = C_d\,A_o\sqrt{\frac{2(P_1 - P_2)}{\rho}}$$

Therefore,

$$\frac{dQ_1}{dt} = \frac{d\left[C_d\,A_o\sqrt{\frac{2(P_1 - P_2)}{\rho}}\right]}{dt}$$

$$\frac{dQ_1}{dt} = \frac{d\left[C_d\,\text{w}\,x_v\sqrt{\frac{2(P_1 - P_2)}{\rho}}\right]}{dt}$$

$$\frac{dQ_1}{dt} = C_d \; w \; \frac{d\left[x_v \sqrt{\frac{2(P_1 - P_2)}{\rho}} \right]}{dt}$$

$$\frac{dQ_1}{dt} = C_d \; w \left[\sqrt{\frac{2(P_1 - P_2)}{\rho}} \frac{dx_v}{dt} + \frac{x_v}{\sqrt{\frac{2(P_1 - P_2)}{\rho}}} \frac{d(P_1 - P_2)}{dt} \right] \qquad (2.44)$$

Using Eq. (2.44) in Eq. (2.43)

$$F_3 = L \; \rho Cd \; w \left[\sqrt{\frac{2(P_1 - P_2)}{\rho}} \frac{dx_v}{dt} + \frac{x_v}{\sqrt{\frac{2(P_1 - P_2)}{\rho}}} \frac{d(P_1 - P_2)}{dt} \right]$$

$$F_3 = \left[c_d \; w \; L\sqrt{2\rho(P_1 - P_2)} \right] \left[\frac{dxv}{dt} \right] + \left[\frac{L \; C_d \; w \; x_v}{\sqrt{\frac{2(P_1 - P_2)}{\rho}}} \right] \left[\frac{d(P_1 - P_2)}{dt} \right] \qquad (2.45)$$

The above equation represents the transient flow force. The first term represents the damping force. The second term in the RHS is of less significance as the pressure rate term does not contribute substantially to valve dynamics and is usually neglected.

When Eqs. (2.45) and (2.41) are combined, they give the complete axial force. The axial force is given by the equation,

$$F_{axial} = 2C_c C_v^2 A_o (P_1 - P_2)\cos\theta + \left[c_d w \; L\sqrt{2\rho(P_1 - P_2)} \right] \left[\frac{dx_v}{dt} \right] \qquad (2.46)$$

2.2.2   Flow Forces on Flapper Valves

Most two stage servo valves have flapper nozzle as their primary or the first stage. This stage provides power amplification. The flapper nozzle has significant forces on the flapper which is mainly due to the static pressure acting on the nozzle area projected on to the flapper. The velocity or dynamic pressure also acts on the flapper and is of interest to include this effect [3].

Let us consider the flapper nozzle configuration as shown in the Fig. 2.3 in which flapper is mounted on a torsion rod with torsion gradient of K_a. This is the case when the flapper is driven from other source such as torque motor where the torque motor armature also serves as a flapper. If the flapper is driven from some other source, K_a can be taken as the equivalent torsion spring constant of the flapper and the driving source.

Fig. 2.3 Double nozzle
flapper valve

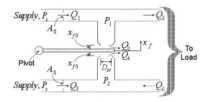

Flow through orifice equation is given by,

$$Q = C_d \ A_o \sqrt{\frac{2(P_1 - P_2)}{\rho}} \tag{2.47}$$

where,

Q is the mass flow rate through an orifice.
C_d is coefficient of discharge.
A_o is the orifice area.
β is the bulk modulus of the fluid used.
P_1 is the pressure at the inlet the orifice.
P_2 is the pressure at the outlet the orifice.
When the flow through orifice is applied across the load, (Fig. 2.4)

$$Q_L = Q_4 - Q_3 = C_{df}\pi \ D_N(x_{fo} + x_f)\sqrt{\frac{2P_2}{\rho}} - C_{do}A_o \sqrt{\frac{2(P_s - P_2)}{\rho}} \tag{2.48}$$

From the Bernoulli's equation we can write,

$$F_1 = \left[P_1 + \frac{\rho U_1^2}{2}\right] A_N \tag{2.49}$$

where U_1 is the fluid velocity at the plane of the nozzle diameter and can be written
as,

$$U_1 = \frac{Q_2}{A_N} = \frac{C_{df}\pi D_N(x_{fo} - x_f)\sqrt{\frac{2P_1}{\rho}}}{\pi D_N^2 / 4} \tag{2.50}$$

Fig. 2.4 Enlarged View of
the Nozzle Section

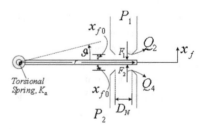

On combining Eqs. (2.49) and (2.50)

$$F_1 = \left[P_1 + \frac{\rho\left(\dfrac{C_{df}\pi D_N(x_{fo}-x_f)\sqrt{\frac{2P_1}{\rho}}}{\pi D_N^2/4}\right)^2}{2} \right] A_N$$

$$F_1 = P_1 \left[1 + \frac{16\,C_{df}^2(x_{fo}-x_f)^2}{D_N^2} \right] A_N \qquad (2.51)$$

$$F_2 = P_2 \left[1 + \frac{16\,C_{df}^2(x_{fo}+x_f)^2}{D_N^2} \right] A_N \qquad (2.52)$$

The net force acting on the flapper is the difference of these forces.

$$F_1 - F_2 = (P_1 - P_2)A_N + 4\pi C_{df}^2[(x_{fo}-x_f)^2 P_1 - (x_{fo}+x_f)^2 P_2] \qquad (2.53)$$

Approximating P_1 and P_2 to be same at null condition or steady state value we get that

$$P_1 \approx P_2 \approx P_s/2 \qquad (2.54)$$

$$F_1 - F_2 = (P_L)A_N + 4\pi C_{df}^2 x_{fo}^2 P_L + 4\pi C_{df}^2 x_f^2 P_L - 8\pi C_{df}^2 x_{fo} P_s x_f \qquad (2.55)$$

A good flapper valve design requires that $x_{fo}/D_N < 1/16$, which makes the second term negligible compared with first term. The third term can also be neglected because $P_L \approx x_f \approx 0$ near null where most operation occurs. Thus the force equation becomes,

$$F_1 - F_2 = (P_L)A_N - 8\pi C_{df}^2 x_{fo} P_s x_f \qquad (2.56)$$

2.3 Transfer Functions of Servo Valves

As mentioned earlier, servo valve systems are highly nonlinear in nature [3]. However, it is found in the literature that nonlinear systems can be satisfactorily represented by linear transfer function models [5] and this is where genius of Fourier and Laplace come handy. The physical phenomenon observed in these nonlinear systems are often described through ordinary differential equations arising from the governing laws and first principles in time domain. Solution to these equations in time domain gets increasingly difficult with increase in variables and solution methodology gets tedious. The solution however becomes simplified when time domain equations are converted into frequency domain using Laplace Transforms. In frequency domain operations, solution methodology involving complex integrals reduces to simple operation of algebraic equations. When these algebraic equations are rearranged, relationship between the system input variable and output variable can be easily established. These input-output relationships for a given dynamic system in frequency domain (or 's' domain) are known as transfer functions.

Though being linearized, a transfer function model should be a close approximation of the real system. The transfer function must be modelled considering all the critical parameters of the real system. The system designer should also have a good vision of behaviour of the system under influence of external influencers like operating conditions and fluctuations. It should be easy to understand intuitively such that a relation between input and output parameters is quickly understood. However, while simplifying the essence of the real physical system should be kept intact. A good model is a judicious trade-off between reality and simplicity. Depending on the simulation outcomes, the system designers can however iteratively increase the complexity of the model until desired results are reached. An important issue in modelling is model validity. Model validation techniques include simulating the model under known input conditions and comparing model output with system output [1].

Several simplifying assumptions should be made to help transform the physical system into a mathematical model. Some of those assumptions, which apply to most of the servo-hydraulic systems, are as follows:

- Energy losses due to bends, fittings and sudden changes in flow cross sections are negligible.
- Sump pressure is negligible and cavitations are not modelled.
- Ports are symmetric circumferentially.
- Servo-hydraulic systems can be accurately modelled using lumped parameters.
- Leakage flows are assumed negligible except for the annular flow for a closed bypass port.

To develop the mathematical models, one has to gather all the information such as what you already know, what assumption you need to make, what are the data's unavailable and most important of all what is the output to be predicted. Once the model is developed we must verify and validate it.

2.3.1 Transfer Function Between the Current and Flapper Rotation

The torque motor consists of an armature mounted on a weak torsion spring, permanent magnets and the armature connected to the flapper. The armature is pivoted and is free to rotate; it is surrounded with the air gap of magnetic field which is created by the permanent magnet. When current is supplied to the armature, the ends of the armature get energized. Depending on the polarity of the current, one end gets attracted and the other gets repelled. This induces a torque on the flapper assembly and causes the armature to rotate about the pivot. The flapper also rotates as it is connected to the armature. The flapper is positioned at equal distances from either of the nozzle, when the flapper gets displaced it blocks one of the nozzles. The flapper obstructs the flow of the fluid and changes the flow balance across the pair of opposing nozzles. The flow rate of the fluid is dependent on the differential pressure, once the flow rate changes, the pressure also varies as other parameters are constant. The variation in the differential pressure between the ends of the spool makes the spool to move in the housing or the sleeve. Thereby opening the orifices and allowing the fluid to pass through orifices and help in actuating the flight control surface.

The movement of the spool forces the ball end of the feedback spring to one side and this induces a restoring torque on the armature/flapper assembly from the spring. When the feedback torque on the flapper spring becomes equal to magnetic forces on the armature the system reaches the equilibrium state, the flapper and armature return to their null position or the neutral position but the spool is in the new position. When the spool is in the new position, it allows the fluid to flow through the open ports.

The total forces acting on the flapper is summation of the forces due to the steady state and transient flow, it also includes the force generated by the differential pressure of the pilot stage acting on the flapper.

The net forces on the flapper is given by, (from Eq. 2.56)

$$(F_1 - F_2) = (P_{Lp})A_N - 8\pi C_{df}^2 x_{fo} P_s x_f$$

The motion of the flapper can be expressed as,

$$T_d = J_a \frac{d^2\theta}{dt^2} + B_a \frac{d\theta}{dt} + K_a\theta + (F_1 - F_2)r \qquad (2.57)$$

where,

T_d is the torque on the flapper.

B_a is the viscous damping coefficient of mechanical armature mounting and load.

K_a is the mechanical torsion spring constant of armature pivot.

J_a is the inertia of the armature and any attached load.

θ is the angle of rotation of flapper.

From the Fig. 2.5,

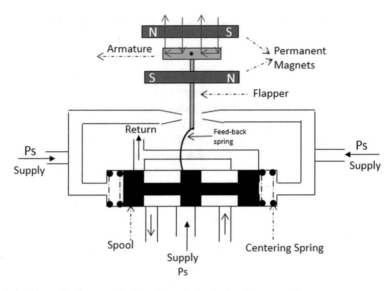

Fig. 2.5 Schematic diagram of a two stage electro-hydraulic servo valve

$$\tan\theta = \frac{x_f}{r} \qquad (2.58)$$

where,

x_f is the flapper displacement

For small angles, $\theta \approx 0$;

$$x_f = r\theta \qquad (2.59)$$

Using Eq. (2.59) in Eq. (2.57)

$$T_d = J_a\frac{d^2\theta}{dt^2} + B_a\frac{d\theta}{dt} + K_a\theta + (P_{Lp}A_N - 8\pi C_{df}^2 x_{fo}P_s r\theta)r \qquad (2.60)$$

The total current developed on the armature due to electrical current input is given by Merritt as [3],

$$T_d = K_t\Delta i + K_m\theta \qquad (2.61)$$

where,

T_d is the torque on the armature due to current.

K_m is the magnetic spring constant of torque motor.

K_t is the torque constant of the torque motor.

Δi is the current input to the servo valve.

Equating Eqs. (2.61) and (2.60)

$$K_t\Delta i = J_a\frac{d^2\theta}{dt^2} + \left\{K_a - Km - 8\pi\ C_{df}^2 x_{fo} P_s r^2\right\}\theta + \left\{P_{Lp}A_N\right\}r + B_a\frac{d\theta}{dt} \quad (2.62)$$

The total deflection of the cantilevered feedback spring at its free end is $[(r + b)\theta + x_v]$, due to the feedback spring the equation of torque becomes,

$$K_t\Delta i = J_a\frac{d^2\theta}{dt^2} + B_a\frac{d\theta}{dt}\left\{K_a - Km - 8\pi C_{df}^2 x_{fo} P_s r^2\right\}\theta + \left\{P_{Lp}A_N\right\}r + \\ (r+b)K_f[(r+b)\theta + x_v] \quad (2.63)$$

where,
   r is the flapper length
   b is spring length.
   x_v is the spool displacement.
   K_f is the spring constant of the cantilevered spring at free end (i.e. spool end)
   This can also be represented as,

$$K_t\Delta i = J_a\frac{d^2\theta}{dt^2} + B_a\frac{d\theta}{dt}K_n\theta + \left\{P_{Lp}A_N\right\}r + (r+b)K_f[(r+b)\theta + x_v] \quad (2.64)$$

where,
   $K_n = K_a - Km - 8\pi C_{df}^2 x_{fo} P_s r^2$ is the net spring rate.
   The transfer function between the current supplied to the servo valve and the angle of rotation of flapper is given by [2],

$$\frac{\theta}{Kt\Delta i - \left\{(P_{Lp}A_N r) + (r+b)K_f x_v\right\}} = \frac{1}{J_a s^2 + B_a s + [K_n + K_f(r+b)^2]} \quad (2.65)$$

2.3.2 Transfer Function Between Flapper Displacement and Spool Valve Displacement

Two stage electro hydraulic servo valves can be classified into three types, based on the feedback method used. The three methods are
   1. Direct position feedback.
   2. Spring feedback to the torque motor.
   3. By placing a stiff spring at the end of the spool.
   The most widely used method is the spring feedback to the torque motor. In this method, a spring is used to convert the position into a force signal which is then fed back to the torque motor. The feedback spring connects between the spool and

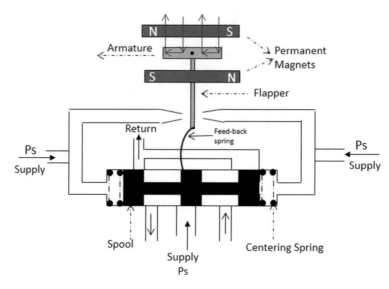

Fig. 2.6 Schematic diagram of a two stage EHSV flapper nozzle spring feedback type

flapper to provide a force balance. The flapper is in turn connected to the armature of the torque motor.

In Fig. 2.6, when there is current difference, the current on the armature creates a magnetic field and this magnetic field interacts with the magnetic field created by the other permanent magnet. This makes one side of the armature to get attracted to one pole and is repelled by the other pole. This magnetic field causes a torque on the armature which rotates the flapper. For instance, if the armature rotates anti-clockwise the flapper moves towards right and closes the nozzle on the right. This increases pressure on the right side and decreases the pressure on left side. This forces the fluid to exert pressure on to the spool and this pressure moves the spool towards left. The spool continues to move to the left until the torque on the flapper due to the feedback spring balances the torque due to input current. Once the feedback spring balances out the torque due to the input current then the flapper comes back to the neutral position or the null position but the spool is in the new position. The new position of the spool is directly proportional to the input current. Assuming linearity about the operating point,

The flow rate across the flapper can be represented as

$$Q_{Lp} = K_{qp}x_f - K_{cp}P_{Lp} \qquad (2.66)$$

where,

Q_{Lp} is the flow rate in the pilot stage.

K_{qp} is the flow gain constant for the pilot stage.

x_f is the flapper displacement.

K_{cp} is the flow pressure coefficient for the pilot stage.

P_{Lp} is the pilot stage load pressure.

The total flow rate across the flapper or the pilot stage is given by,

$$Q_{Lp} = A_v \frac{dx_v}{dt} + \frac{V_0}{2\beta} \frac{dP_{Lp}}{dt} \qquad (2.67)$$

where,

A_v is the end area of the spool.

V_0 is the volume of the fluid in the nozzle.

β is the bulk modulus of elasticity of the fluid.

x_v is the spool displacement.

The force acting on the spool is the force due the mass of the spool and the steady state flow forces, hence we get;

$$A_v P_{Lp} = M_v \frac{d^2 x_v}{dt^2} + 0.43w(P_s - P_L)x_v$$

where,

M_v is the mass of the spool, w is the area gradient.

$$P_{Lp} = \frac{M_v \frac{d^2 x_v}{dt^2}}{A_v} + \frac{0.43w(P_s - P_L)x_v}{A_v} \qquad (2.68)$$

Equating Eqs. (2.66) and (2.67),

$$K_{qp}x_f - K_{cp}P_{Lp} = A_v \frac{dx_v}{dt} + \frac{V_0}{2\beta} \frac{dP_{Lp}}{dt} \qquad (2.69)$$

Using Eq. (2.68) in Eq. (2.69)

$$K_{qp}x_f - K_{cp} \left[\frac{M_v \frac{d^2 x_v}{dt^2}}{A_v} + \frac{0.43w(P_s - P_L)x_v}{A_v} \right]$$

$$= A_v \frac{dx_v}{dt} + \frac{V_0}{2\beta} \frac{d\left[\frac{M_v \frac{d^2 x_v}{dt^2}}{A_v} + \frac{0.43w(P_s - P_L)x_v}{A_v} \right]}{dt} \qquad (2.70)$$

On expanding,

$$K_{qp}x_f - \frac{K_{cp}M_v}{A_v}\frac{d^2x_v}{dt^2} - \frac{0.43wK_{cp}(P_s - P_L)x_v}{A_v}$$
$$= A_v\frac{dx_v}{dt} + \frac{V_0\,M_v}{2\beta\,A_v}\frac{d\left[\frac{d^2x_v}{dt^2} + \frac{0.43w(P_s-P_L)x_v}{A_v}\right]}{dt} \tag{2.71}$$

$$K_{qp}x_f = \frac{V_0\,M_v}{2\beta\,A_v}\frac{d^3x_v}{dt^3} + \frac{K_{cp}M_v}{A_v}\frac{d^2x_v}{dt^2} + A_v\frac{dx_v}{dt} + \frac{V_0\,M_v}{2\beta\,A_v}0.43w(P_s - P_L)\frac{dx_v}{dt}$$
$$+ \frac{0.43wK_{cp}(P_s - P_L)x_v}{A_v} \tag{2.72}$$

Dividing A_v on both sides.

$$\frac{K_{qp}x_f}{A_v} = \frac{V_0\,M_v}{2\beta\,A_v^2}\frac{d^3x_v}{dt^3} + \frac{K_{cp}M_v}{A_v^2}\frac{d^2x_v}{dt^2} + \frac{dx_v}{dt} + \frac{V_0\,M_v}{2\beta\,A_v^2}0.43w(P_s - P_L)\frac{dx_v}{dt}$$
$$+ \frac{0.43wK_{cp}(P_s - P_L)x_v}{A_v^2} \tag{2.73}$$

Let us take x_v common,

$$\frac{K_{qp}x_f}{A_v} = \left[\frac{V_0\,M_v}{2\beta\,A_v^2}s^3 + \frac{K_{cp}M_v}{A_v^2}s^2 + s + \frac{V_0\,M_v}{2\beta\,A_v^2}0.43w(P_s - P_L)s + \frac{0.43wK_{cp}(P_s - P_L)}{A_v^2}\right]x_v$$

$$\frac{K_{qp}x_f}{A_v} = \left[\left[\frac{V_0\,M_v}{2\beta\,A_v^2}\right]s^3 + \left[\frac{K_{cp}M_v}{A_v^2}\right]s^2 + \left[\frac{V_0\,M_v}{2\beta\,A_v^2}0.43w(P_s - P_L) + 1\right]s + \left[\frac{0.43wK_{cp}(P_s - P_L)}{A_v^2}\right]\right]x_v$$

Dividing the equation by the constant term and renaming it to be ω_f,

$$\omega_f = \frac{0.43K_{cp}w(P_s - P_L)}{A_v^2}$$

$$\frac{K_{qp}x_f}{A_v}\frac{1}{\omega_f} = \left[\left[\frac{V_0\,M_v}{2\beta\,A_v^2}\frac{1}{\omega_f}\right]s^3 + \left[\frac{K_{cp}M_v}{A_v^2}\frac{1}{\omega_f}\right]s^2 + \left[\frac{V_0\,M_v}{2\beta\,A_v^2}0.43w(P_s - P_L)\frac{1}{\omega_f} + \frac{1}{\omega_f}\right]s + 1\right]x_v$$

$$\frac{x_v}{x_f} = \frac{\frac{K_{qp}}{A_v}\frac{1}{\omega_f}}{\left[\left[\frac{V_0\,M_v}{2\beta\,A_v^2}\frac{1}{\omega_f}\right]s^3 + \left[\frac{K_{cp}M_v}{A_v^2}\frac{1}{\omega_f}\right]s^2 + \left[\frac{V_0\,M_v}{2\beta\,A_v^2}0.43w(P_s - P_L)\frac{1}{\omega_f} + \frac{1}{\omega_f}\right]s + 1\right]} \tag{2.74}$$

References

1. A. Maria, Introduction to modelling and simulation, in *Proceedings of the 1997 Winter Simulation Conference, NY, USA* (1997)
2. M. Jelali, A. Kroll, *Hydraulic Servo Systems—Modelling, Identification and Control* (Springer, 2003)
3. H.E. Merritt, *Hydraulic Control Systems* (Wiley, New York, 1967)
4. T.P. Neal, Performance Estimation For Electrohydraulic Control Systems, in *The National Conference on Fluid Power in Philadelphia*, Pennsylvania. Moog, NY, Technical Bulletin 126 (1974)
5. W.J. Thayer, *Transfer Function for Moog Servo Valves* Moog Technical Bulletin 103

Chapter 3
Servo Valve Characteristic Curves

Typically, a servo valve is characterized by a series of curves which provide behavioural information of the servo valve under varying operating conditions. These curves are typically used in identifying the valve centre type, null bias, response to inputs of varying amplitudes at varying frequency and phase difference between system input and output at different operating frequencies.

The type of valve centre is defined by the width of the land compared to the width of the port in the valve sleeve when the spool is in neutral position. If the width of the land is smaller than the port, the valve is said to have an open centre or to be under-lapped, as shown in Fig. 3.1. A critical-centre or zero-lapped valve has a land width identical to the port width. A valve with a land width greater than the port width is called closed centre or overlapped.

3.1 Flow Characteristic Curve

The flow characteristics of the valve may be directly related to the type of valve centre. Corresponding to Fig. 2.6, there are three important flow gain characteristics, with the shape shown in Fig. 3.2. In fact, it is better to define the type of valve centre from the shape of the flow gain near neutral position than from geometrical considerations. A critical centre valve may be defined as the geometrical fit required to achieve a linear flow gain in the vicinity of neutral position, which usually necessitates a slight overlap to offset the effect of radial clearance.

The flow characteristics of the valve may be directly related to the type of valve centre. Corresponding to Fig. 2.6, there are three important flow gain characteristics, with the shape shown in Fig. 3.2. In fact, it is better to define the type of valve centre from the shape of the flow gain near neutral position than from geometrical considerations. A critical centre valve may be defined as the geometrical fit required achieving a linear flow gain in the vicinity of neutral position. This usually necessitates a slight overlap to offset the effect of radial clearance.

© The Author(s) 2019
J. J. Vyas et al., *Electro-Hydraulic Actuation Systems*, SpringerBriefs in Applied Sciences and Technology, https://doi.org/10.1007/978-981-13-2547-2_3

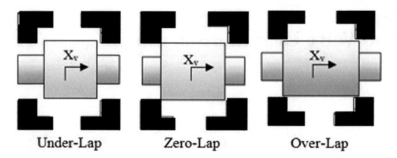

Under-Lap   Zero-Lap   Over-Lap

Fig. 3.1 Different valve lapping when the spool is in neutral position

Fig. 3.2 Flow gain (load
flow, QL versus spool stroke,
xv) of different centre types

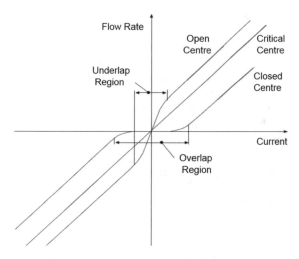

A majority of four-way servo valves are manufactured with a critical centre because of the emphasis on the linear flow gain. Closed centre valves are not desirable because of the dead band characteristics in the flow gain. With a proportional amplifier, the dead band results in steady-state error and can cause backlash which may lead to stability problems in the servo loop. It is possible to compensate for dead band electronically but it will at least influence the response time of the servo valve.

Open centre valves are used in applications which require a continuous flow to maintain an acceptable fluid temperature and/or an increase of the hydraulic damping. However, the large power loss in neutral position, the decrease in flow gain outside the under-lap region and the decreased pressure sensitivity of open centre valves restrict their use to special applications.

Thus, with the objective of experimentally identifying the valve centre type, a ramped input current from 0 to +20 mA and 0 to −20 mA was commanded to the flapper nozzle servo valve at a constant supply pressure of 70 bar. It is to be noted

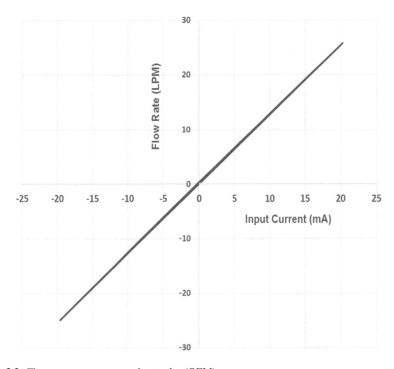

Fig. 3.3 Flow rate versus current input plot (OEM)

here that, since the nominal pressure of the valve is defined at 70 bar, this test needs to be necessarily carried out at 70 bar (Figs. 3.3 and 3.4).

From the flow through orifice equation,

$$Q = C_d x_v w \sqrt{\frac{2(P_s - P_L)}{\rho}}$$

where C_d is the flow discharge coefficient of the servo valve, x_v is the spool displacement, w is the land width, ρ is the oil density, P_s is the supply pressure and P_L is the load pressure. Since the outlet ports to the servo valve are directly routed to the tank, the load pressure P_L is equal to zero. Hence, with all the other terms remaining constant, the outlet flow from the servo valve is directly dependent on the spool displacement and the supply pressure. To test this formulation, the ramped input current to the servo valve was commanded at different supply pressures. As the supply pressure increased, the servo valve took lesser current input to completely open. This observation is presented in Fig. 3.5.

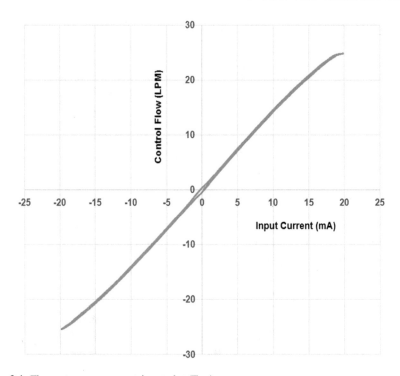

Fig. 3.4 Flow rate versus current input plot (Test)

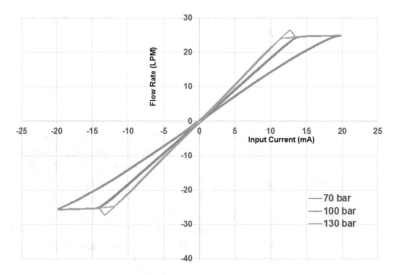

Fig. 3.5 Flow rate versus input current at different supply pressures

3.1.1   Bode Plot

As mentioned earlier in Chap. 2, it is convenient for servo system analysis or any other studies concerning system synthesis to first represent the system by a simplified transfer function. Such transfer functions are at best, an only close approximation of a physical servo valve. While it is true that representing a physical system through a transfer function will result in loss of non-linear characteristics, the useful and approximation capabilities of linearized transfer functions for servo valve response in analytical work is well established. Further, linearizing the system also avoids the complex task of solving ODEs leading to convoluted integrals. This implicates that ODEs, when linearized, become algebraic equations which are much easier to solve. However, for satisfactorily close approximations, it must be taken care that many design factors and other operational and environmental variables are duly addressed. Otherwise, the simulated results produce significant differences with respect to the actual dynamic response. Considering the critical parameters of the valve design, it is shown clearly in Chap. 2 that internal valve parameters (e.g. nozzle diameter, orifice sizes, spring stiffness coefficient, spool diameter, total internal spool displacement, etc.) may have a great influence in the dynamic response of the servo valve. Once a servo valve is manufactured, the dynamic response will depend greatly on the operating conditions like supply pressure, commanding signal amplitude level, hydraulic fluid operating temperature, valve resistive loading and so forth. These effects are minimal while operating in the close vicinity of the design values, but should be considered where high deviations are anticipated. It is important to appreciate and control these parameters and other operational variables when performing measurements of servo valve dynamics. The transfer function approach will thus enable a system designer to incorporate all the important parameters into the model and also provide good means to solve it by linearization. This directly facilitates analysis of system behaviour across a broad spectrum of input amplitude and input frequency using the famous Bode plot technique.

   Bode plot, a logarithmic plot, is one of the easy and helpful tools in defining the system behaviour. The plot made from the transfer function gives the frequency-domain response of any dynamic system. The gain (in dB) and phase (in degrees) margin obtained from the magnitude and phase values plotted over frequency, f, respectively, can be directly used in identifying the stability of the system. The gain margin is the amount of gain increase or decrease required to make the loop gain unity at the gain crossover frequency, Wgm where the phase angle is $-180°$. The phase margin is the difference between the phase of the response and $-180°$ when the loop gain is 1.0. The output magnitude is given by

$$A = 20 * \log_{10}(|A_i|)$$

   From the gain and phase frequencies of the system, the stability can be identified as follows:

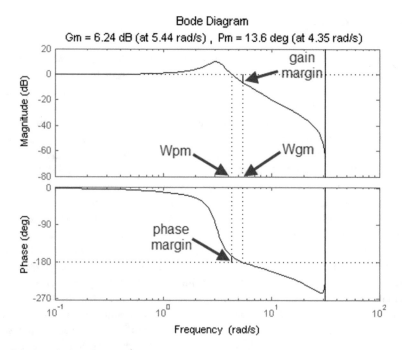

Fig. 3.6 A typical Bode plot

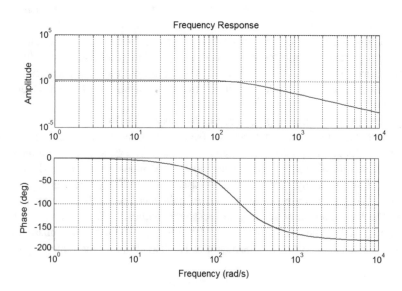

Fig. 3.7 Bode plot from system identification

(1) When the phase margin is greater than the gain margin, then the system is defined as the stable system.
(2) When both gain and phase margin are equal, then the system is identified to be marginally stable.
(3) If the phase margin is lesser than the gain margin, then the system is termed as unstable system.

The typical Bode plot below shows the gain and phase margin (Figs. 3.6 and 3.7):

The Bode plot presented above has been built through system identification toolbox in Matlab. The procedure involved has been detailed in Chap. 4.

Chapter 4
System Identification

System identification is the process of developing mathematical models for the dynamic system of interest, which can be used for the modelling and simulation, prediction, control design, error detection, etc. The process involves the various experimental procedures carried out for variable input signals, followed by the identification of mathematical model that best fits the system of interest. The accuracy of the selected mathematical model is then validated using System Identification Toolbox/MATLAB. The model with the best fit can be used for the further analysis of the system.

The general purpose of the system identification is carried out to

(1) Predict the future response of the system,
(2) Tuning or designing of the controller,
(3) Analysis of the system,
(4) Developing the existing model in order to optimize the system design.

The application domain of system identification started expanding with the evolution of the tools and methods available for the prediction of future responses, improvising the control system design, etc. The numerous fields that use the system identification technique include electrical or electronic systems, chemical process, biology, hydrology, economy, environment, psychology, biomedical research, etc. Models can be formulated either by using first principle or data-driven technique.

4.1 Modelling

Modelling of a system is done either with the knowledge of physical laws, i.e. first-principle approach, or with the experimental data, i.e. data-driven approach. The first-principle approach is a time-consuming and complex exercise as compared to the latter. The experimental data obtained cannot be used directly for the model development. The data should be checked for various factors such as noise, outliers

© The Author(s) 2019
J. J. Vyas et al., *Electro-Hydraulic Actuation Systems*, SpringerBriefs in Applied Sciences and Technology, https://doi.org/10.1007/978-981-13-2547-2_4

and missing data. Outliers can be defined as the data, which deviates from the others data points due to sensor defects or abrupt changes. The detection of outliers is very complicated as there is no universal reason. The missing data can be due to sensor malfunctioning, data transfer losses, sampling rate nonuniformity, etc., the quality of the data can be verified by data visualization through which the outliers, drifts or any other changes can be identified. Other information on the process such as delay, gain, and insight into dynamics can also be seen. Once the analysis of the experimental data is carried out, the modelling of the process can be initiated with the data available.

The identification of the model is carried out in two steps, namely,

1. Specification of the structure and order of the model,
2. Estimation of the model parameters.

The model selection is considered as the most challenging part of identification. The following steps are carried out to validate the model for the intended use.

1. The variance and bias of the model can be checked with the estimation algorithm.
2. The prediction accuracy of the model can be defined with the help of the n-step ahead prediction.
3. The model should be capable of serving the purpose of the intended use.
4. The prior knowledge on the type of model is required for the selection of model.

Once the model is selected for the required purpose, optimization of the model is carried out in order to achieve further prediction accuracy with a further minimization of prediction errors. The model assessment step is an integral part of the model development step. The ability of the model in effectively explaining the output variations is studied. The only goal of performing the quality assessment is to ensure the prediction error is minimized to the extent possible.

For a good working model, the following suggestions must be followed. (1) data quality must be good enough, (2) visual analysis should be carried out, as it is a qualitative analysis, (3) model selected should be based on the intended use, (4) complex models should be avoided, (5) data should be collected for the appropriate timescale, (6) model validation to ensure the quality of the model.

The models can be classified based on their dependence on the physical laws and formulations into three types, namely,

(1) White Box Model,
(2) Grey Box Model,
(3) Black Box Model.

White box models are the models that reflect the transparency of the process, which can be explained by a mathematical form of the model through the physical laws that are present. The degree of transparency increases with the information available to describe the process. This model comes under the category of first principle technique.

Though the information is available on the system, not all parameters are known, so this type of model is termed as *Grey box modelling*. This model can be used to estimate the parameters that are unknown in the system through system identification. This model is also known as the semi-physical model, as it involves both the first principle and data-driven technique.

The models, which use the data-driven technique, without any transparency about the process and do not depend on any physical laws involved in the process are termed as *Black box models*. This model can be used when no prior information is available on the system. This type of models provides better accuracy when compared to first principle methods by careful analysis of the data and identification procedure.

4.2 System Identification in MATLAB

The system identification is performed using MATLAB's System Identification Toolbox, Garnier et al. (2003); Ljung (2009); Young and Jakeman (1980). The toolbox has a useful graphical user interface (GUI) which is used to import the frequency/time domain data from the lab test results acquired through NI data acquisition card. The toolbox is used to estimate time-continuous transfer functions from time response data, which should characterize the dynamics between the reference current and the measured flow response. An overview of the system identification toolbox is shown in Figs. 4.1 and 4.2.

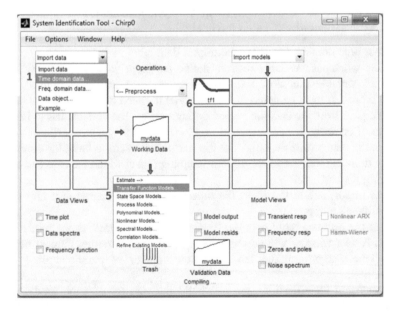

Fig. 4.1 System identification toolbox—GUI

Fig. 4.2 Test data import GUI

Following are the steps involved in the system identification process using MATLAB's system identification toolbox.

1. Select time domain data from the import data tab
2. In the workspace variable section, define the input variable (i.e. current) and output variable (i.e. flow rate) in the exact same names as defined in the workspace. It must be noted that these variables need to fed into the MATLAB workspace from the data file logged during testing. The variable has to be in the form of an array.
3. Give the data a suitable name. Set the start time to zero to build the continuous time domain models and specify the sampling rate at which the data was logged during testing.
4. Press import button to import the time domain experimental data into system identification app.
5. From the estimate tab, select the 'transfer function model' as the model structure. A new window will pop up where the number of zeroes, number of poles, prediction algorithm and the maximum number of iterations need to be specified.
6. The system identification app will build the transfer function model of the specified order. Upon double-clicking the built model, the transfer function

model can be accessed. Other tools like the model output, transient response, frequency response will present the curve fitting accuracy, the step input response and the bode plot of the system, respectively.

Chapter 5
Results and Discussion

It is noted from the literature that though the servo valve system nonlinear is often and satisfactorily modelled as a linear system. The system model is represented by a second-order transfer function [2] relating the servo valve output flow rate to the servo valve input current. The model is represented as

$$\frac{Q}{i} = \frac{K}{\frac{S^2}{\omega_n^2} + \frac{2\zeta S}{\omega_n} + 1} \tag{5.1}$$

where ω_n is the natural frequency of the system, K is the servo valve flow gain and ζ is the damping ratio. However, working from the first principles, the relation between the input current and output flow rate is shown in equation below. This equation will be derived in the succeeding section.

$$\frac{Q_L}{K_t \Delta i - K_w x_v} = \frac{K_3 K_2 \frac{1}{K_f}}{A_v \cdot s \left(\frac{S^2}{\omega_n^2} + \frac{2\zeta S}{\omega_n} + 1 \right)} \cdot$$

To identify parameters given in Eq. 5.1, a ramp input current varying from +20 to −20 mA is commanded to the servo valve and the output flow is recorded. The plot for Q versus I is shown in Fig. 3.3. The plot indicates negligible dead band. If the dead band were to exist, the valve can be tested for a current input with 50% positive and negative offset and current amplitude equal to 5, 10 and 20% of maximum current. This input will steer clear the dead band region and the system response analysis in each case can be carried out using frequency response plots. However, since the servo valve in our case has a negligible dead band, the input current and the output flow rate are directly used to fix second-order transfer function model in system identification toolbox on MATLAB.

© The Author(s) 2019
J. J. Vyas et al., *Electro-Hydraulic Actuation Systems*, SpringerBriefs in Applied
Sciences and Technology, https://doi.org/10.1007/978-981-13-2547-2_5

5.1 Grey Box Modelling

MATLAB offers System Identification App with a user-friendly GUI to import time/frequency domain data and build mathematical relations between input and output variables. In order to carry out system identification, the modelling of the input excitation signal plays a vital role. An input signal with extremely high perturbations and variations results in the poor fits and extremely simple input signals will result in the poor mapping of the dynamic characteristics. Different models of input signals are detailed in [1]. For our purpose, two models of input signal namely the pseudo-random signal and chirp input (0.5–2 Hz) with the maximum amplitude of ±20 mA is selected. In each case, the time domain data of input current in mA and flow rate in LPM are imported into the system identification app and a second-order transfer function model is built.

5.1.1 Test Case 1: Current Signal—Pseudo-Random Multiple Signal (PRMS)

The pseudo-random current input signal and the corresponding flow rate output is shown in Fig. 5.1. The transfer function model (built on system identification app, MATLAB) with 2 poles and 0 zero is found to predict the test results with the best possible accuracy of 81.4% and the model with 2 poles and 1 zero is found to

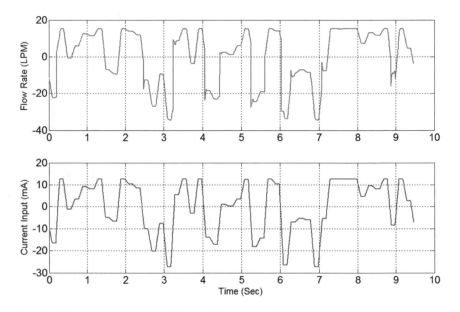

Fig. 5.1 Measured output (control flow) and input signal (current input)

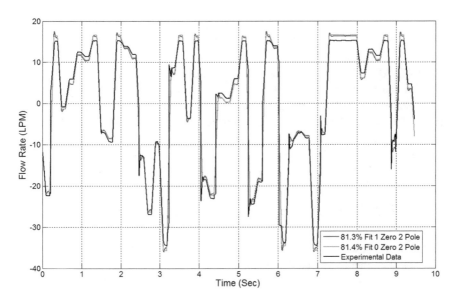

Fig. 5.2 Measured and simulated output of PRMS input

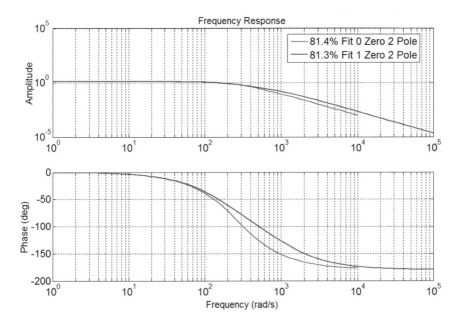

Fig. 5.3 Frequency response plot of PRMS

predict the test results with an accuracy of 81.3%. The magnitude plot and the phase difference plot for both the models are presented in Fig. 5.3. The transfer function predicted using PRMS as excitation signal is as follows:

$$\frac{Q}{i} = \frac{217400}{s^2 + 1114s + 166300}$$

MATLAB uses inbuilt algorithms like the predictor-corrector method, the classical least squares method, etc., to validate the models. A closely approximated model has the best fit with the experimental data. From Fig. 5.2, it can be seen that, among the two models, the maximum fit possible was 81.4%. This is considered to be poor in experimental identification. Hence, the input current signal is modified to a chirp signal (Fig. 5.3).

5.1.2   Test Case 2: Current Signal—Chirp Signal

The chirp input current signal and the corresponding output flowrate is shown in Fig. 5.4. The predicted transfer function models from the chirp signal input produced favourable results with 95.4% being the best fit case with 2 poles and 0 zero and another model with 2 poles and 1 zero predicted 89.3% fit. Hence, the model with 2 poles and 0 zero was selected for parameter identification. The measured versus simulated model output and the frequency response plots for the test case (2) are shown in Figs. 5.5 and 5.6.

The transfer function consisting of 2 poles and 0 zero built from system identification is given as

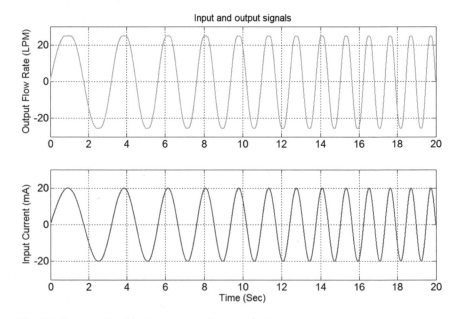

Fig. 5.4 Test case 2—chirp input–output flow rate and input current

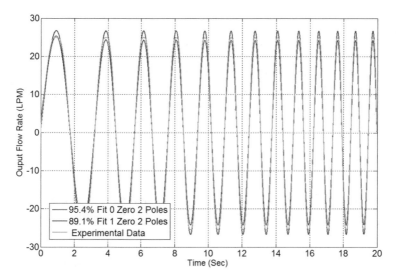

Fig. 5.5 Measured and simulated model output—test case 2

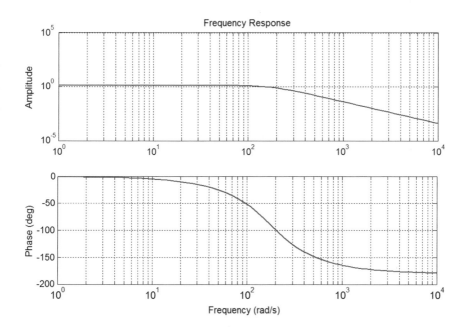

Fig. 5.6 Frequency response plot—test case 2

$$\frac{Q}{i} = \frac{14567.29}{s^2 + 147.83s + 11449.86} \tag{5.2}$$

Comparing Eq. 5.2 with Eq. 5.1 and also from the bode plot, the natural frequency ω_n is approximately 132 rad/sec, system flow gain K equals 1.272 LPM/mA and the damping ratio is approximately 0.69.

5.2 White Box Modeling

In this section, a relation between the output flow rate and input current for a given flapper nozzle-type servo valve is derived from first principles. The derivations follow 'Hydraulic Control Systems' [2] by HE Merritt and other standard references.

A simplified construction of the flapper nozzle valve is shown in Fig. 5.7. Torque applied at the armature pivot point due to input current is given by

$$T_{Applied} = K_t \Delta i + K_m \theta \tag{5.3}$$

where K_t is the Torque Motor Gain (in-lbs/mA), i is the input current (mA), K_m is the magnetic spring constant of the torque motor (in-lbs/rad) and θ is armature deflection (rad).

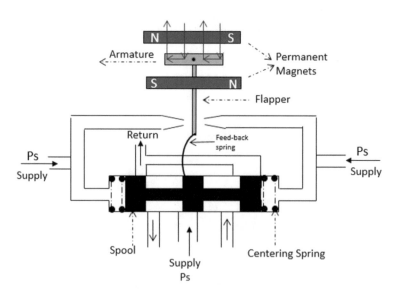

Fig. 5.7 Flapper nozzle valve—construction

The resistive torque to armature deflection is given by

$$T_{Resistive} = J_a\ddot{\theta} + B_a\dot{\theta} + K_a\theta + T_L \tag{5.4}$$

where J_a is the rotational mass of the armature (in-lbs/in/sec^2), B_a is the net armature damping (in-lbs/in/sec), K_a is the mechanical torsion spring constant (in-lbs/rad) and T_L is the load torque due to feedback wire. The load torque can be given by

$$T_L = K_w x_v \tag{5.5}$$

where K_w is the feedback spring constant (in-lbs/in) and x_v is the spool displacement. Equating Eqs. (5.4) and (5.5), we get

$$K_t\Delta i + K_m\theta = J_a\ddot{\theta} + B_a\dot{\theta} + K_a\theta + K_w x_v \tag{5.6}$$

Considering $K_a - K_m = K_f$ as the net stiffness of the flapper armature assembly and Laplace transforming Eq. (5.6)

$$\frac{\theta(s)}{K_t\Delta i - K_w x_v} = \frac{\frac{1}{K_f}}{\frac{s^2}{\omega_n^2} + \frac{2\zeta s}{\omega_n} + 1} \tag{5.7}$$

where $\omega_n = \sqrt{\frac{K_f}{J_a}}$ is the pilot stage natural frequency and $\zeta = \frac{1}{2}\frac{B_a}{K_f}\omega_n$ is the pilot stage damping ratio. Multiplying Eq. (5.7) with a constant r (the distance between armature pivot and pole face), we get x_f the flapper displacement.

Now, the hydraulic amplifier differential flow to the spool stage ΔQ is given by

$$\Delta Q = K_2 * x_f \tag{5.8}$$

where K_2 is the hydraulic amplifier flow gain (in^3/sec/in) and thus multiplying Eq. (5.7) by 'r' and further multiplying with K_2 will yield the following:

$$\frac{\Delta Q(s)}{K_t\Delta i - K_w x_v} = \frac{K_2\frac{1}{K_f}r}{\frac{s^2}{\omega_n^2} + \frac{2\zeta s}{\omega_n} + 1} \tag{5.9}$$

At the spool stage

$$\Delta Q = A_v\dot{x}_v \tag{5.10}$$

where x_v is the spool velocity (in/sec). Laplace transforming the above equation

$$\Delta Q(s) = A_v \cdot s \, x_v(s) \tag{5.11}$$

Substituting Eq. (5.11) in Eq. (5.9):

$$\frac{x_v(s)}{K_t \Delta i - K_w x_v} = \frac{K_2 \frac{1}{K_f} r}{A_v \cdot s \left(\frac{s^2}{\omega_n^2} + \frac{2\zeta s}{\omega_n} + 1 \right)} \tag{5.12}$$

Considering K_3 as the spool stage flow gain constant (in^3/sec/in), the output flow rate is given by

$$Q_L = K_3 x_v \tag{5.13}$$

Substituting Eq. (5.13) in Eq. (5.12), we get

$$\frac{Q_L}{K_t \Delta i - K_w x_v} = \frac{K_3 K_2 \frac{1}{K_f} r}{A_v \cdot s \left(\frac{s^2}{\omega_n^2} + \frac{2\zeta s}{\omega_n} + 1 \right)} \tag{5.14}$$

These set of relations from Eq. (5.3) to (5.14) are modelled on the Simulink platform and is compared with the grey box model as shown in Fig. 5.8, (Table 5.1).

As seen in Fig. 5.9, there exists a very small discrepancy in the predicted values of flow rate in case of both grey box and white box models. The servo valve rated flow at ±20 mA is 25 LPM. The grey box model and white box model are predicting approximately 25.6 and 26 LPM, respectively. The error of less than 5% is within an acceptable margin. The system identification is an iterative process and more precise modelling of the input signal, calibration of sensors, the rate of sampling etc., will be vital to minimizing the error.

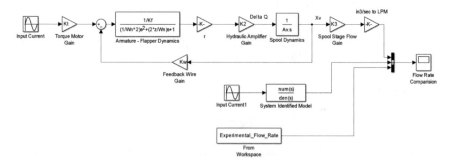

Fig. 5.8 Simulation of white box model and comparison with grey box model and test results

Table 5.1 Simulation parameters

Parameter	Value	Unit
K_t	0.0249	in-lbs/mA
K_w	16	in-lbs/in
K_f	92	In-lbs/rad
K_2	112	in^3/sec/in
K_3	920	in^3/sec/in
ω_n	735	Hz

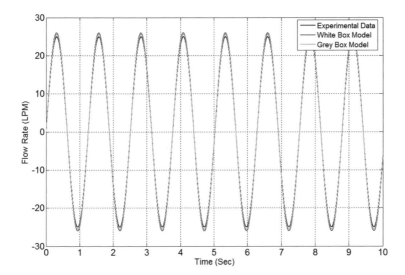

Fig. 5.9 Flow rate comparison

References

1. M. Jelali, A. Kroll, *Hydraulic Servo Systems—Modelling, Identification and Control* (Springer, 2003)
2. H.E. Merritt, *Hydraulic Control Systems* (Wiley, New York, 1967)

Chapter 6
Conclusion

This brief highlights the importance and the necessity of carrying out system identification in the context of electro-hydraulic actuation system for flight control applications. The advantages of system identification in building real-time plant models of the physical systems and its subsequent application in dynamic simulations and virtual prototyping have been highlighted. A servo valve model has been built using test results and the system identification toolbox in Matlab. A physical setup consisting of flapper nozzle type servo valve and a double acting actuator connected to loading system has been used for this purpose. The servo valve model is useful in the flight control design optimization and can result in reduced design cycle time and reduced procurement costs. The servo valve model built from system identification and the analytical white box model are simulated in Matlab and the results are compared with experimental data.

From the results, the mathematical model obtained from the grey-box model gives the best fit of 95.4% for the 0.5–2 Hz chirp signal. The second-order transfer function model (2 poles and 0 zero) obtained through this method is used for the further identification of system parameters such as servo valve gain and natural frequency. The simulation model and the test data from the test rig show that the accuracy of the model identified with the help of measured data is satisfying. Finally, the tests carried out to compare the performance plot from supplier also have a satisfying match between the measured data and the supplier plot. Therefore, it is concluded that system identification can be a very handy tool when it is intended to build a mathematical model of dynamic systems like that of hydraulic systems.

© The Author(s) 2019
J. J. Vyas et al., *Electro-Hydraulic Actuation Systems*, SpringerBriefs in Applied
Sciences and Technology, https://doi.org/10.1007/978-981-13-2547-2_6

Printed in the United States
By Bookmasters